3

W9-ADD-129

FOLLOWING
GRAVITY

THE VIRGINIA COMMONWEALTH UNIVERSITY SERIES
FOR CONTEMPORARY POETRY   WALTON BEACHAM, GENERAL EDITOR

Moving Out
  by David Walker,   1976

The Ventriloquist
  by Robert Huff,   1977

Rites of Strangers
  by Phyllis Janowitz, 1978

James Cook in Search of Terra Incognita
  by Jeanne Larsen,   1979

PS
3551
.P67
F6

James Applewhite

FOLLOWING
GRAVITY

University Press of Virginia   *Charlottesville*

Salem Academy and College
Gramley Library
Winston-Salem, N.C. 27108

THE UNIVERSITY PRESS OF VIRGINIA
Copyright © 1980 by the Rector and Visitors
of the University of Virginia

First published 1980

Library of Congress Cataloging in Publication Data

Applewhite, James.
    Following gravity.

    I. Title.
PS3551.P67F6 811'.54 80-21578 ISBN 0-8139-0885-X
Printed in the United States of America

For Jan

FOREWORD

For some writers a subject matter is a form of destiny. Whether James Applewhite, taking thought, chose his native South as subject or whether he has become its witness largely by virtue of birth and residence may come to the same thing in the end, and yet I doubt that it was ever for him a question of choice. Not that Applewhite's poetry, except in the most general way, sounds very much like that of other southern poets. There is, for instance, a scarcity of the bombast and extravagance associated with certain southern tendencies in literature as well as politics. But, like the best of southern writing, Applewhite's is at heart traditional, and all the more resonant for being so.

It is a simple matter to respect a body of work like this. I think I first began to feel something beyond respect—something, in my case, like affection—on noticing the care this poet had taken to name the very names of his neighbors, whether, as in some cases, drinking or, more soberly, attending funerals. I was not beyond thinking, rather grandly, of Homer naming the Achaian chieftains or Milton the fallen angels. For there is a drift or pull in Applewhite's poetry toward myth and legend, a mark of high ambition which his tact and reserve normally keep from view.

But before the myth comes the fact, and Applewhite is first of all a realist. His proper nouns are not the easy satiric devices they have become in much contemporary writing but arise instead from a feeling for the authentic, for the fact of what is or has been. Memory is involved, and a wish to record and keep things. Because his subject is so often the humble and the ordinary the style may be allowed a degree of elevation, even a certain composed and formal character, suggesting mannerliness and a sense of occasion. Now and then a passage will come through with something of the effect of a photograph, a photograph that in the development stage has been perhaps a little treated in the dark room. Yet always the fact counts for as much as the art.

The present book represents a continuation and an extension of the work begun in Applewhite's first collection, *Statues of the Grass*, and a confirmation that it had indeed come, as the best work does, deeply from character.

DONALD JUSTICE

ACKNOWLEDGMENTS

The author and publisher wish to thank the following magazines for permission to reprint certain poems:

Poetry for "Firewood" and "The Waters" copyright 1976 by The Modern Poetry Association, reprinted by permission of the editor.

New England Review for "Elephant Graveyard." Reprinted by permission from *New England Review*, Vol. II, No. 2 (Winter 1979). Copyright © 1979 by James Applewhite.

The Atlantic Monthly for "Tobacco Men."

Harvard Magazine for "Pamlico River," first published in *Harvard Magazine*, 81, no. 6 (1979).

Southern Poetry Review for "My Grandmother's Life."

Poetry, Now for "Drinking Music" and "The Mary Tapes."

Carolina Quarterly for "Some Words for Fall" and "Following Gravity" (originally "Following the River").

The South Carolina Review for "From as Far Away as Dying."

Mississippi Review for "The Call," "January Farmhouse," "Autumn Ivy," "Building in the Country," and "Blood Ties: For Jan."

Poets in the South for "Afternoon Stream" and "After Your Birthday."

St. Andrews Review for "Grandfather Noah."

Vanderbilt Poetry Review for "Elegy For a General Store."

The author wishes to express his gratitude to the John Simon Guggenheim Memorial Foundation for its generous support.

CONTENTS

I

Tobacco Men

Late fall finishes the season for marketing:
Auctioneers babble to growers and buyers.
Pickups convoy on half-flat tires, tobacco
Piled in burlap sheets, like heaped-up bedding
When sharecropper families move on in November.

No one remembers the casualties
Of July's fighting against time in the sun.
Boys bent double for sand lugs, bowed
Like worshippers before the fertilized stalks.
The rubber-plant leaves glared savagely as idols.

It is I, who fled such fields, who must
Reckon up losses: Walter fallen out from heat,
Bud Powell nimble along rows as a scatback
But too light by September, L. G. who hoisted up a tractor
To prove he was better, while mud hid his feet—
I've lost them in a shimmer that makes the rows move crooked.

Wainwright welded the wagons, weighed three
Hundred pounds, and is dead. Rabbit was mechanic
When not drunk, and Arthur best ever at curing.
Good old boys together—maybe all three still there,
Drinking in a barn, their moonshine clearer than air
Under fall sky impenetrable as a stone named for azure.

I search for your faces in relation
To a tobacco stalk I can see,
One fountain of up-rounding leaf.
It looms, expanding, like an oak.
Your faces form fruit where branches are forking.
Like the slow-motion explosion of a thunderhead,
It is sucking the horizon to a bruise.

A cloud's high forehead wears ice.

Snow over Soil

Shoveling the light, ephemeral loam
Of snow, I uncover an earlier season:
My father is spading us a foxhole
Deep as his crotch, for our pretended war.
Old brown Doc and his mule are splitting
And folding back the wild onion sod.
He mops his brow, looks at his brass pocket watch,
Leans on his plow across from my father.
Past the pecan uprooted in a storm to come,
A dirt street paved now for years,
The farmer in the corner of a field chops wood.

Soil is falling inward in round-cornered squares.
The plowshare moves like the fin of a shark
Patrolling underneath our garden. Days have
Rotted at the edges like old plank fences.
These three figures wheel together through
The unfenced fields, triangle
Space of our yard between them.
My father digs a foxhole for that earlier war.

Grass flashes bright, sun-burnished, like Doc's
Brass watch, whose time hasn't held.
The ax of the man across a field
Comes down on a log. A *chunk* strikes my ears
So surprisingly late,
It seems set free from the stroke which fell,
Keeps ringing through the walls of days
Which let light through like this vanishing loam.

Drinking Music

I

Cornstalks and arrowheads, wood rot cleanly as wash
Hung in wind. They are covered by this roof as I think:
The losers, the fallen, the kind who go under, faces
As familiar as Civil War casualties to the soil's imagination.
Whiskey workers with no front teeth, men from down home,
With leather boots made by the *Georgia Giant*, denim or khaki
Their only other clothes, the dye scrubbed away
By brush in fields and seasons of laundry
So a sand white cotton shines through.

They stand for the showing of true colors, on soil
Instilled with the sun's going down, which crests up
At evening in a brick hue of broom sedge, clots below
In sweet potatoes succulent as blood.

Their ears give over to a jukebox excuse poured
Slow as molasses. Hank Williams' whiskey forgiveness.
The twang has a body like tenderloin and turnips.
Paul Junior Taylor, with your Budweiser belly,
Your shoulders muscled wide like a tackle's,
I wonder what it's done with you now.

II

The broom sedge fields were ruddy from sunset.
Cold whiskey is the color of straw.
Sky ain't hardly kept no color at all,
And Lord I'm lonely for the ground.
I'm far from home but near.
I'm high, so high and low. Sweet chariot.

The song is red, like what men eat.
Sky is clear as the ninety proof shine they drink.
Lord, Lord, a man is a funny piece of meat.
Their boots print fields that understand their feet.

Elephant Graveyard

A wreckage we recognize. By a Ford
    like father with spectacles, hear how silk
over buttocks whistles still in a seat, use
    your empty heart to locate that one
point of blood at the virgin beginning:
    when reveries shone perfect as corneas, before
wishing solidified to Pontiacs, Kennedy-era
    Cadillacs smashed head-on to Camelot.

Sift with your nails under seats, through
    the buckshot pebbles, cigarettes' tinsel,
pearl buttons from back-road adulteries.
    See sparrows flicker profiles of Lincolns,
their rear view mirrors aimed a hundred ways
    in afterthought. Birds in low sun
mirror bits of this day, this air, bear
    light which makes these artifacts legible.

Elegy for a General Store

Sweet potatoes the color of terra cotta
    lay binned beside coal. Twine on
cones spooled long enough for summer,

remembering already the work at a tie bench,
    where tobacco leaves were caught up
into "hands" thick as fish, then looped. Nails

clicked, slickly as nickels, when we hand-fulled
    them sharp into a sack. Steel bullets,
little pencils, they held the plans of houses

in their shanks. Chains linked for timbering,
    hooks crooked stubbornly as the pull of a tractor,
cables coiled ready to come alive with strain,

slender as dynamite, more dangerous than snakes,
    when trembling with an engine. Saws
with the teeth of sharks were not all,

there were spirit levels, squares—as perpendicularly
    just as Old Testament elders—planes to skim
slivers into curls like paper. Window panes

with white paper tags seemed to stare already
    as at clay rises, broom sedge, and pines.
Flypaper was sticky to gum buzzes asleep

where front porch railings barred the sun.
    Straw hats stacked beside shoes stored
a shadow under brims, to receive wives' faces.

Miles away from anywhere, with Christmas
    bringing tinsel and tea sets, only birthdays
left to come, they lived unmusically, sang

in churches made empty by sun and the whiteness,
    accompanied by organs like voices from the grave.
Whether snakes poisoned joy, or rifles

from those shelves killed snake, hawk, and joy,
    my grandfather's store is deserted. Houses
which turned toward roads their soda-cracker faces

are contained no longer in levels. The plan
    for that strictness, which survived while oaks
grew intricate with their hundred seasons,

is lost. No salt for hog killing is sold.
    No jars for canning the tomatoes or relishes.
Yams like bulbs for some giant iris

hid from their planting a kernel of sweetness.
    Their scriptural society was separated by distance
and denial. Voices beside upright pianos wore

Salem Academy and College
Gramley Library
Winston-Salem, N.C. 27108

hoarse with sweat from those brows, made harsh
      by Adam's apple. That clay never flowered.
Sermons had withered up the glands for singing,

calico smothered the orchids which unfolded
      in new wives. Hymns traveled yards,
rubbed over roots like snakes frozen still.

Thin as the telephone wires strung for miles,
      or a spent bullet's whistling into distance,
nails seemed links in loneliness' iron chain.

Echoing within walls of houses going up,
      they protested a wooden hollowness. Windowbox
petunias would prove shallow, sashes grow blind

with pure blue, where a shotgun had knocked into
      memory one hawk-wing crucifix. Hammered in,
they shrieked like being crowbarred out of houses.

The Call

Enormous nostalgia, call of unconsciousness,
You remember a geography south of where
The Mississippi flows. Clouds pile, generic
But unique, individual as always. Honeysuckle
Scent sheathes streams, along hollows
Below waking. Mourning doves call,
From a continent drowned with Atlantis.
A child across the street is fretful from heat.
I hear her paradise and the beginning of weeping.

Building in the Country

These fields toward the river are outside time.
The horse has been grazing forever.
That black mass of pines
Is such as must always exist. My straight
Route to Carolina Builders leaves out of account
Those leaves in peripheral vision, which hold
Their benediction just beyond focus.
I envy dead farmers their roofs, their rust:
At home in these spaces. Ripples
Of motion where a dove alighted
Undulate in meadows like seacoast breakers.

Friday afternoon, the house going well,
My contractor and I surrender:
Unmatchable beauty of mountain laurel,
Accumulated atmospheres, farmhouses' shaded-over wells.
We drive Orange County's backcountry,
Getting drunk in a red Ford pickup,
See a golden eagle rising high,
Over almost the spot where his bootlegging father
Hid the still in a chicken house.
The hundred dollar bills
He remembers piled up on the kitchen table
Blew away like leaves when Pa died of cancer.
We look through an honorable old house
By the river, windows broken out

Under eighty-year oaks.
Kicking at beer cans, I see that our families
Leave headstones for legacies.
Why try to homestead an undertow of grass?

Maybe our misery gives that finishing perspective,
As when barns hold hills like a varnished canvas.
Atmosphere gathers most deeply to the wells of these yards,
Most breathlessly recedes
Where a house-corner cuts it:
Canopied by oaks, suspended on distance.

Water

Downward in direction as a willow,
    white rapids flowering in your hair,
O life-and-death lingering
    and beginning, element of Ophelia:

your communities are rotifers
    churning; puddles cupped in stumps
round the womb-sacs of insects,
    hold daphnia, flutters of cilia—

yet you are an expanding glitter,
    wind scaling bronze across a river,
are puddles with pollen and frogs
    where mud bubbles the noise rejoice.

O pattered-down patterning,
    prefiguring of spray upon a bough,
you are liquid for spring's generation,
    for tadpole sperm to gather in.

Your molecules link into chains
    slippery for the capillary thirst
which trees satisfy, defying gravity,
    though entropy drains you toward the sea.

Material spirit, vapor in breath,
      gauze for our entrance and exits,
you sheathe our wet begetting,
      are breathed out finally at death.

You always are single yet double:
      where oxygen triangles, the one
between two, with androgynous hydrogen,
      a woman bends supple.

II

My Grandmother's Life

I

Edges of the pane are scented with steam.
One pot is boiling on the cast-iron range, thin lid
Falling *tink, tink,* after each bubble.
My grandmother is peeling peaches into an enameled pan.
She slices the sweeter, bruise-darkened flesh onto
Spots chipped black like knotholes. I know all
This but see only through the window,
Which is part of myself. I look down on
A garden within a border of bricks, in shade of a pecan,
Which rocks that ground like a floor of water
When wind rises in it. I see sage and thyme and parsley.
In possessing those markings, bricks in soil
Strewed white from our sandpiles, I possess my life.

Grandfather's strength stands in chunks of wood
By the range, pine bark still on some
Like frosting on cake. A five-gallon carboy
Smells of the water of Seven Springs.
Half an hour away, it starts by the river
That the creek of our town curls into.
That space is filled with the balloon-head musings
Of clouds and trees, with a mourning dove seeming
To whistle all back toward birth with the timbre of its cry.

Mineral odor from the heavy glass jar
Seems surrounding the roses, clasps as with a skin
Those rheumatic knottings twined into time.

Down the lane out of view, past the packhouse
Furred with splinters, ranked with lilies,
Where grasshoppers roosted, chewing tobacco,
To be caught by their cellophane wings,
I recede into time, toward a granite slab
Under the scion pecan of my uncle's planting.

On that stone seat cool in its shadow,
I am a child with my velvet cousin,
A boy with the tomboy climbing our tree.
I roam away, feel the other loves layered in pain
On the love which alone is like breath and sight,
Am tossed by winds like the pecan.

My grandmother's presence, permanent as water,
Shapes itself about leaves and roses
Enclosed by a square made of bricks.

II

My grandparents had rested together in their graves
For years. Their house was rented to strangers.
I lived in a city, listened to traffic.
Visiting my parents, I had found my grandfather's carboy,
Had washed it and set off driving, to feel my way,
Through the sandy farms, again to Seven Springs.

Fields lay low within breath of the river,
The yards overflowing with flowers:
Daisies, alyssum, sweet william, petals in facets
To separate the sunlight like prisms.
The Neuse slid still in its bends with a rippled sheen.
I drove beckoned to by wisps of its vapor,
The highway in distance cracked open
By wedges of the Spanish moss crowding over.

Back in the city, restless with fever, I experienced
The trip in my head. Turning to my wife,
I was embracing her belly, her thighs,
Again in memory searching for the spring.

Late afternoon now, thunderheads piled
In giant indifference on their one-way mirror,
Repeating one another like headstones' marble.
Road openings winked between pines and sweet gums:
Wrong ways, deceitful seduction.
My wife lay beside me, her thighs
In curves like the beach beside a bay,
Burning me like sunlight.
Another turn in the road and a hollow in trees
Opened out into a lane, the path wound a hill,
Past the old spa's wooden hotel,
A dip in the slope and dust broke open
Upon a springhouse rich with water.
I had touched the mouth of her womb.

I rose to a light through shrubs across
Our window: some lonely bulb or a star.
A dream shone in mind
From the month of my grandmother's death.

We are wading in a bay.
A yellow beach encloses it like arms,
Like the bell of a flower.
I feel myself retracing steps.
As she leads me by the hand, her figure, slight,
Slender as a girl's, is again a girl's.
In straits between distant sand,
Bay and sky arc together.
My baptism awaits in a shimmer of light.

III

Stopping for Gas

Entering a small town (Sims or Baily,
A crossroads) we see and smell the lumber stacked
Crosswise to dry, sawed lengths from hearts
Of trees the color of sunlight.
In an Esso station's expression,
With its greasehouse going into shadows,
I recognize the face of a memory:
When Esso red on gas pumps and pillars
Showed vacant as the sun's afterimage.

Nine or ten years ago. I pulled to the pumps
And stood out eaten by my thoughts.
Bought a bottled orange drink to suck
Like sawdust in the water-shimmer heat.
I walked on wobbly knees to the john,
Let my stream rattle down like dried peas
Into the basin: that day I was leaving Durham
And my wife who had first left me
For good I thought if a sick numb
Thing in my head and body like an iron
Truck tire wedge was a thought.

Now on these trips to visit my parents,
Sometimes I stop here for gas.
I buy us orange drinks
As if really from trees, while the big
Country kid in coveralls works the pump,
I see and smell cut pine like my wife's
Blond hair, take my son to the Men's room where
We cross our streams like a magic charm.

Rooster's Station

Our Esso station faced its mirror-image station,
Where red-haired Rooster sold the beer Father hated.
As wings gathered soft around our bulbs, blacks
From the vague dirt streets passed live through our light:
Glad-colored as moths or long-feathered birds.
Their shirts' blood-crimsons, gold or green blouses,
Sang within my eye like birdcall. They gathered
In flocks to Rooster's, buying Budweiser and Miller's,
Blue Ribbon and Schlitz. Stragglers
Now and then fell to us for Pepsis,
Or a brown boy exactly my height wanted money
Of pleasure, the gold foil condoms I dispensed
From the high shelf box with an envious knowledge.
Sundays, seated among our Puritan colors,
I remembered the blouses like tropical birds;
A sympathy perched with clipped wings within me,
Or fluttered with the scarlet-shirted
Bloods crossing out of our lights,
Into creekwater Saturday dark across fields,
Pockets enriched by the gold coins I sold.

Blood Ties: For Jan

In the forty-nine Ford I had earned,
We printed our tires
Into dirt below our special tree.
Ascending on wings
Of insects and leaves, we flew
In a slippage that was tight
And all freedom
On a wind above
Corn in the dark.

Our blood in an updraft column
Lifted off
In a suspended moment
Out of touch with the earth
As if endless
Like a fountain evanescent under lights
Till it burst
Toward the stars
Like our bonfire sparks at night

And then drifting
Winking out
Until lightning bugs hovering
In a honeysuckle odor
When we raised our heads to the window again
Were scattered over fields
Like embers of burning
As if we'd seeded the air with our fire.

Our bodies retreated,
Blood flowing west toward the long-gone sun.
We lay back happy
But hushed, mute and grave,
Afraid of the permanence of the love we had made,
Afraid of our knowledge: two after Eden.

Pamlico River

I breathed that odor of land-draining water,
Leachings from ditches and saw-bladed marshes,
From springs, field-trickles, now channeled by creeks
Into a five-mile flood turned bronze in the sun:
Cypresses ever in the distances, living
And dead, fish hawks nesting their skeletons.
I breathed that odor of ending and beginning,
Land's drift marrying with salt and the tides.
I lay on a spit of sand in the sun,
Savoring the taste of my body and water.
My cousin Ethel cooked steaks on a fire,
Ethel's beau and I sipped beer. That spirit
From childhood, whose cloud-imagination
Trailed the rain in necklaces, felt winelike
Arteries and veins, intoxicating stems,
Like grandfather's scuppernong: grapes in leaves
Grown yellow with October too sweet to resist.

January Farmhouse

Snow on ground and
Brown weeds above: patches
Like fragments of dinner plate
Where sun brushes clay.
The washboard wall is in shadow,
Holds skim milk light
The way a bedsheet hung out to dry
And catch cold's cleanliness
Gathers sheen from the sky.
The white boards appear
Translucent, like a woman's skin
When she is old and left alone
The January afternoon;
Seem translucent with enclosing
Light I see through an upstairs window
Collected in a dresser mirror;
Or see from glimpsing
Through front and back windows,
All the way through those rooms,
Through this still afternoon
In her life and back into sky,
Where sun slants invisibly
Without clay, broom sedge,
Or features to make rosy, there
Where wind's too thin to be seen.

White Lake

Rimmed in by cypresses, tin water flashed
Like the top of a can, in fields still buzzing
With cicadas: electrical August short-circuiting.
The surface slicked over us like oil, shone
Silver with clouds. We walked, holding hands,
Toward the rides: Roller Coaster. Dive Bomber.
We sat in the Ferris Wheel, throbbing
With its engine, as it hurried us backward,
To show a black polish, the lake like marble
Under stars, bulbs on its opposite shore
Rolling across reflection in miniature pearls.

With a wince of thrill in the quick of our spines,
We offered up ourselves to a turning as enormous
As the seasons or desire, whirled down to search
Shadows, where water lapped subtly at roots,
For a place we could lie down together, wandered
Through glare from the lighted piers,
Till we took our chances below a capsized boat.
In the rides park afterward, there were dolls
To be won by rings or thrown balls.
Pandas, like drunken guests at a wedding,
Formal in their black and white, faced
A tree-tall whirling as if spun by a giant.

Afternoon Stream

The river runs under,
Through noon's moving glitter
Standing still.
I make water by the river,
Feel sexual encounters in multitudes,
Lizards of darkness,
Slip through that mind
Of its own
In my hand. The mind
Seemed a thing made of light,
Unmoved by slippages
Into
And from under.
I thought that it was noon,
I wanted it marriage
And beauty
But it is struggle
For mastery,
I said that it was noon . . .
It is later.

Autumn Ivy

After the clouds and a quarrel,
We walked where light
Across a slope broke free,
To pick out, in particular,
One tree.
A pine stood gloomy as the night,
But encircling leaves resembled
The lip of a candle,
Translucent, as if a fire
Were shining through: that trunk
Held rigid in a serpentine
Embrace, a three-leaf pattern—
Poison and forbidden—
Yet so rosy a yellow the vine
Seemed for grapes, as if curled
From Eden or the underworld.
On a slope between sun
And that darker expanse,
Weighing how blood ran bitter
And loving, we walked on happier,
Keeping our distance.

Grandfather Noah

The cemetery slope, grass after frost,
Looked pale as their hair. Earth's swells rolled,
An ocean of uncertain footing. My townspeople
Walked it to pay last respects, Nell Overman's
Round hat against the horizon, above her drawn-back
Shoulders, Norwood Whitley rubbing his blue jaw,
Planting steps heavily as if in boots to the knee,
Ann Louise Stanton moving soberly in her frock,
These Christians fallible across wintry grass
Which called them down, there to face the dying
Of W. H. Applewhite, gold name on a window,
Though the general store was his own no longer;
Who'd lived with the sun, the returns of plowing,
His devotion inscribed in two rows of his garden.
I imagined his figure, bowed to the handplow,
Old man half-kneeling within the order of seasons.

The people still moved in a stream, as I had seen
Them down steps of our church, Wainwright's round head,
His one crippled shoulder turning that hand inward,
My community as individually weathered as the trees,
Coming from inside, where I had sat in their power,
While flowers opened upward with notes of the organ.
The bronze coffin, magnified by the intense

Eyes of that full congregation, seemed massive,
Oblong, as I had imagined King Arthur's barge.

Borne up the aisle on townsmen's shoulders,
It seemed the vessel of hereditary identity,
Profoundly measured, onto seas past knowing.
I watched through the window as earth fell in waves,
The bronze box burning in my thoughts, an ark
Of my covenant, ark of this succeeding Noah,
King whom I would ride with on water till doomsday.

After Your Birthday: For Jan

It is not only the moment,
It is years of your lips,
The slip
Inside you of what is "my body"
In speech, but now a sensation
Of wet things I've seen and touched
Rising within you, my country:
Thunderhead at evening
In the horizon of autumn,
And all summer long.

The look of your eyes
Is deepened by histories.
Your breath holds that night
Beneath our private tree,
Its canopy
A cloud against stars, when
We offered each other
The necessary fruit.

Now you are forty
But still my new land,
Breasts I rise from a child again;
You are water for the river
Of everything I know,
The flow in a bed I must follow.
Together
We will run to the sea.

IV

Firewood

After the axhead has flashed
And the maple log is sawed,
He lingers on the hearth,
Anticipating light, for this sun

In wood is somehow in his blood,
As his eyes flicker clearly
Their spark in the thicket
Of a world not understood.

It is not only a golden
Living descended to wood
That the child's struck match
Frees to dancing,

There are October's odors
Veined into foliage, which a boy
And his man of flame
Exhale as smoke to the cold.

This story he thinks, a blond
Prince lost in a forest,
Is as tragic and old
As a chemical formula.

Not only fire descended
To water and fiber, but wonder
At the union he senses,
That to the earth's antennas

Of living branches, sun should
Signal through clearness,
To maple leaves and apples,
Coloring with sugar,

For a meeting more than either.
He senses how flame returned
From leaves toward sun
Musks air with mortal October.

Following Gravity

I

Surface is the place things change
     if at all. Clouds come
alive from inverted horizons.

Pine and oak and their movement
     in wind are narrowed into
air's one dimension, dragonflies

in sun in the running mirror
     a focus of four wings—
picture from the nineteenth century.

A brown face slicks over snags,
     shines oily as sweat. Men
hewed barns, now color water coffee.

Heat above willows, a permanent
     medium, glazes estates under
hills of trees. Where a creek comes in,

windows have the white country look,
     house I'd imagine for a cousin.
I met her at a family reunion.

Her face summed a roomful of Barneses—
        grandmother's people—noses bridged
high into the brow, lips vertically

lined to be stern. But those Roman
        features were lovely with her eyes:
collective ancestor, distant sister.

A child had fallen from a tree.
        Her pupils enlarged with his pain.
I imagined the molds we spill from.

        II

Heat is a varnish, has hardened
        into air's one dimension. Clouds'
bloom rises from a horizon of pines.

Sunlight seats her in front, skinned
        with gold like a statue of Athena.
The canoe follows on toward recollection.

As two drops fall from my paddle,
        mind steps away from its ribs,
crosses water. Quartz makes the steps.

Morning glories twine the split rails.
        A room shows white tongue-and-groove.
Two windows watch as the river bends.

They see through the distance and haze
      as if over a horizon. Beside her,
I breathe in a morning-glory color.

On glass shelves, trinkets
      reflect in each pane. Metal
grenadiers, curious bottles.

She says I should touch a blue dish,
      with maple leaf inside. My kite
trails its loop of string away,

seems a yellow leaf on water.
      My fingers are dreaming of cornhusks.
She suggests I handle a clay pot.

Broom sedge feathers inside it.
      I walk toward home with my brother,
our green canoe left shadowed by a bridge.

Her nail touches shell in the clay,
      touches mother-of-pearl.
Moon sheens the mallard we carried.

III

Are things arranged here before
      the times they capture? I've had
sensations like remembering the future.

She had me see shells on her shelves,
      like forms for our feelings. The mussel
grew into them in water we'd canoed.

For minutes made ready by mother-
      of-pearl, the river seemed birth.
Eddies made whorls like fingerprints.

If these are the patterns for days, where
      are others? She smiled at the question.
Cases held figures on all walls:

cast-metal cannon, cameos. Doors upon
      doors, toward still farther rooms.
Where is my death? came aloud.

She lifted a sphere full of water.
      Many quartz marbles seemed
returned there: drops to a fountain.

Where's the life I'll never have
      with you? She led to the kitchen,
working the handle of a pump, said

Drink. My face bit a tumble of water.
      Its odor was ocean, distance
and iron. I drifted the river.

IV

Between one bend and another,
        a chapter of history. Corn lined
the fields, each blade like a sickle.

Her voice was confused by the willows.
        Springs seemed rotting into motion
one unfissured light, like granite.

The river was eroding it, into particles
        of sand to make loam, its silt
envisaging the women, dead men,

battlefields, tears underwater, earthy blood.
        Without its ooze, golden-skinned summer
would be a single day and never end.

Legs and breasts you've touched, hands
        quick as fish, live only
in this autumn stream. I know,

I said, paddling in the horsefly
        heat. *I don't accept your argument.*
I want you, for myself, someday.

But corn still tasseled, and streams
        breathed seasons from their hollows.
Even wind seemed following gravity.

Some Words for Fall

The tobacco's long put in. Whiffs of it curing
Are a memory that rustles the sweet gums.
Pete and Joe paid out, maybe two weeks ago.
The way their hard hands hook a bottle of Pepsi Cola,
It always makes me lonesome for something more.
The language they speak is things to eat.
Barbeque's smell shines blue in the wind.
Titles of Nehi Grape, Doctor Pepper, are nailed
Onto barns, into wood sides silvered and alive,
Like the color pork turns in heat over ashes.

I wish I could step through the horizon's frame
Into a hand-hewn dirt-floored room.
People down home in Eastern N.C.,
When they have that unlimited longing,
They smell the packhouse leavings.
They look at leaves like red enamel paint
On soft drink signs by the side of the road
That drunks in desperation have shot full of holes.
No words they have are enough.
Sky in rags between riverbank trees
Pieces the torn banner of a heroic name.

From as Far Away as Dying

And now in the end I can see this community together,
Under angles of poor wooden gables and porches,
Accumulated in vision and fronting the west at evening,
With figures fixed in humblest gesture, descending
A warped step, arising from porch swing or rocker,
Or stooping to spit tobacco, become an architectural face
Above Salisbury's entrance, man and wife in cotton
Rosy with sun as if King at Wells with elbows
Thrust from his throne, or stone-wimpled Lady medievally
Distant in that air I remember, like choirs of all souls:
Beside posts, porch railings—voices, this saintly communion.

V

The Mary Tapes

I

Mary Woods was my name, before I married
A Hill. The recorder was Roger's present,
So now I've got to think what to say.
My mind is on the house we're building,
And Raleigh seems so far from our town down home,
I feel like I've moved to a different century.
But those first days don't leave you. I remember
Living there, in a tenant shack out behind grandma's.
Sun made the roof tin creak like a stove
Cooling off; our hall felt rickety, pine boards
That bounced you up and down when you walked.
Near the kitchen stove, where the linoleum ended,
I could see white flashings through the cracks,
When our chickens ran underneath the floor.

I'm halfway embarrassed to hear myself talk,
When it comes back out so flat and country.
I can manage their English well enough—
Yes, my dear, I see your point perfectly.
But the way you talk when you're being *correct*
Won't let my aching get eased into words.
This loneliness catches in my throat,
And tears seem beading like a sweat inside.
Their regular words go numb and deadly.
When I try to think of who I really am,

I see a circle of water in a mason jar.
I'm carrying it to daddy, where he's plowing tobacco.
Dirt through the toes of my sandals feels powdery
And hot, like wood ashes shoveled from the stove.
It seems old-timey, with daddy in the field,
And him and the mule kind of small and so lost
In the corner by the river, where trees billow up
High as a cliff, and thunderheads are piled even higher,
Like an avalanche ready to bury us in whiteness.

Listening to a mourning dove call, I'd forget
What I was doing, like I had lost my own name.
My eyes might as well have been from
Somebody dead, a woman out of one of those yellow
Tin pictures, her dress draped around her like curtains.
A sadness would be running with the river from horses,
That crippled up their riders, or carried them to battle.
And the water would be sliding inside a smell
Like a skin, snaking away the sunlight.

What I was lonesome for felt close by then.
My great-great-grandparents and aunts and uncles
Were there in that field but hidden out of sight,
And the ones who were swept away and drowned
When they tried to cross the ford in high water
Were still in a stream running underneath the river.

People can't ever be happy. The folks
Down home were ignorant and ill as rattlesnakes.
Families would quarrel, or talk would start
About a woman and the preacher, till her husband
Had his shotgun out, with his eyes like he'd drunk
From a kerosene lantern and was about to catch fire.
I have to remember the loneliness and meanness,
Along with the feelings of forever. We'd swim
In a pond, and come up with water on our lashes,
So the bushes and grass would look early with dew,
Like drying off from just being made.
I didn't feel then like I ever could die,
Or bad things ever could happen. The way
That air felt in the fields, like biting an apple,
Was everything I wanted. Having Jimmy scared me,
But it was the sadness later that changed my feelings.
Jimmy was just a baby in his crib.
I'd walked up the road, to see if there was mail.
When I came back in, he was blue and still.
The covers were not over his face or anything.
It near about killed me. I cried for a week
Like I was crazy. I wouldn't eat meals
Or look in people's faces, just sat with my fingers
Kind of messed up in tears and in my hair.
I found out later that crib deaths happen.
I took it on my own self then.
The one thing that saved our marriage, was Roger
Never said even the least little word to blame me.

I couldn't stand snakes run over on the pavement,
Or frogs squashed flat to little skins.
It worried me when wind piled high in the clouds,
And swooshed against our house with rain in a river.
Lightning looked as crooked as Satan's pitchfork.
Everything that was dirty and evil seemed
Hopping and crawling out of bushes and ditches
When storms were making veils on the fields.
I was the one who had let it all happen,
Had let Roger get in between my legs
When we were still in high school, parked on
The ox road at night, when honeysuckle breaths
From the ditches would tangle inside with our panting.
That was why Jimmy had died, I imagined.
So we sold our little piece of the farm,
And Roger went to school at N.C. State.

Here in the Piedmont, you don't feel so ignorant
And doomed, with planes taking off for New York,
And hospitals and doctors if the kids get sick.
Ginger and Cindy are seven and nine,
And I'm thirty-seven. Now we've decided
We like the country after all. The land
Up here isn't scary, with its rolling hills;
The sky doesn't crowd so close at the horizon.
Some radio tower or T.V. antenna
Is poking up there in the distance, and the trees
Don't look too powerful, so you love them better.
I especially like falls, with daisies and chicory

By the road, and sweet gums turning their different colors.
Magazines like *Better Homes and Gardens*
Gave me a lot of ideas for our house.
Our lot is in a woodsy new section,
With a stream over rocks along the back.
A drawing of exactly the right roof for the place
Will shine in my head, the rafters raised up
Like sunbeams, and marking out a space in the trees.
The plan we're building is almost like that.
Our stream has pools that are clear to the bottom,
With minnows there fanning tame as goldfish,
And leaves crumpled up on the surface, drifting
With the wind, as if not really touching the water.
Trees on our lot will be fresh as the woods
Down home, but with briars cleaned out, and poison
Ivy vines cut from next to the rocks.
Our furniture, and books in their glass-fronted cases,
Will make our rooms as bright as in a picture,
So shadow from a crow passing over, or a snake,
Can't make us feel gloomy enough to die.

                    II

Well. It may be just a dream. But things do
Get easier inside me, the more I understand.
Putting in tobacco, I'd laugh with the others,
But sometimes I'd feel uneasy, my bosoms
Little budding-up things, like halves of green apples.
The woman named Nank, who helped us, was quick

As a mink, could strike out dangerous as a moccasin.
The razor blade between her two fingers
Had already cut you when you saw it.
Her Indian cheekbones made her eyes look squinty.
Field niggers leaned in the packhouse door
Where she was grading at the table. It gave me
Goose bumps, she looked so womanly dangerous,
Her bosoms in that shirt showing deep between.
The dust looked brassy in sun through the door,
Like the summer stored up. The hand wanting Nank
Was afraid to be sassy, he knew she'd cut him
Like a cold wind had carried a razor.
Her eyes looked as black as two buttons,
And after he was gone, a sick-feeling
Wonder and longing was left in my stomach.
Nank would tickle so it hurt my bosoms
And ask me if boys had ever done it to me.
I felt all flushed like hot water inside.
Grown-ups' secrets could cramp you like colic.
When a woman at the revival got crazy
With Jesus, and her lips turned back,
Like grinning and hurting at one time,
I imagined how her loving must have felt.

Sometimes mamma reminded me of Nank,
Her blond hair kinked up on her head,
And blue eyes that burned you clear through,
Like church windows shining on a sinner.
She made it so hard for daddy to be right.

She acted as if men were overgrown boys,
Like daddy'd just throw away our money
If she let him keep account of it himself.
I remember him standing on a pier at White Lake,
A beer in his fist, looking at the girls
In tight shorts. "Just because you've ordered, don't mean
You can't look at the menu," was what he said.
He couldn't stand the same thing himself.
When daddy got jealous, momma hadn't done
A damn thing, except get cold against him.
She liked to get me over on her side,
And kind of shut him out of our lives.
It seemed like the house must be shivering from wind,
The way they were arguing. I turned hot and cold
In bed from their voices back and forth,
Till a slap sounded out, and a screaming crying.
Then sometimes later I heard them screwing.
Loving and hating are mixed up in folks,
But surely to God I can understand it better
With Roger than my folks in that house down home.
I hated that summertime dreading, when the heat
Ponded up, and people were hitting or grabbing
In their soaked-through shirts, looking drunk,
Or asleep, or like walking underwater,
Till you expected for their hair to float up.
The only explanation was a preacher's
"Damnations" and "hells" rumbling in a revival,
Muttering together with the bullfrogs and thunder.

People up here have air-conditioned homes.
The one we're building has insulated walls,
So you can't hear talk from the room next door.
Roger will have a desk, and our furniture
Will be safe behind the upstairs windows,
So the river can't drown me in that helpless feeling.
Things seem more reasonable here in the Piedmont.
When Roger got a mind to chase tail, I didn't
Let it feel like the end of the world,
Like people down home, when they'd kill somebody.
We met this doctor and his wife at a party,
And he was a pretty-faced man, the kind
Whose wife is real plain, about as good-looking
As he might be, with a wig and some powder.
So anyway I made up this complaint, with heart
Palpitations, and dizziness when I first stood up.
The appointments were set in the late afternoons,
And after I got out, I'd shop, or drive around.
Roger looked worried when I got home
One evening after nine. I asked him
If the kids had had supper. He said they all
Went to Hardee's, when I didn't come back.
The appointments were all late, I said.
His pupils looked dark and as big as an owl's.
He asked me to tell if that doctor was
"Abusing my trust." I laughed in his face.
I wasn't Snow White, or his daughter, I said.
If he wanted to be chasing some narrow-assed girl,
Who just barely knew enough to type,

Don't talk to me about some sacred trust.
And if he had a taste for new dishes,
Maybe I would eat out myself.
He grabbed me right rough by the shoulders,
And shook me so the hair fell in my face,
And my hand got loose by itself and slapped him,
And he caught me by the throat, with his eyes
Like they were staring into too bright sun
And didn't even see me. I asked if he loved me,
Because folks down home used to kill the ones
They loved. His face changed quick, like a cloud
Before the sun. We sat down trembling,
And talked it out, and afterwards made love.
I'd wanted to get away from all that,
But took after my folks enough, I guess,
To have to have a love-fight with Roger.
I can't blame them too much, remembering
What we came up from. Daddy got started
With wiring work during World War Two,
On a hanger for blimps in Elizabeth City.
Later he climbed poles in the middle of the country,
And worked up to be the superintendent
Of a whole town's lights and water.
You think we weren't proud, in our shiny Buick,
Driving down the main street at night, when store windows
And cars looked kind of famous with lights,
When my daddy was the one in charge of the current.

I'd always stuck to daddy like a burr,
Before I got to realizing how different girls are.
Mamma first made me doubt him, I guess.
She said that girls had to watch out for boys,
They had this thing like a snake where ours
Was just a crack, and could get it inside you.
She said a man's business was as ugly as the skin
On a bat's wing. *I* don't think they're ugly.
They rear up heroic and full of blood,
Like a horse on hind legs, stiff enough
To screw the whole world, then afterwards are shriveled
Like a piece of dough or putty. They're funny
More than anything else, and make me feel
Kind of protective and forgiving. But when men
Walk swaggery, like they were waving and wobbling
It around right then, it doesn't make sense.
A man's just half of the loving. But when a bunch
Get hard into their work, and bulge big muscles,
With their leather belts full of a lineman's tools,
You'd think they could *fuck* the whole world.
If a woman comes by, they yell like savages.
Men don't really understand. They bluster
And bully and want to do a sneaky thing to you,
Then get away. *Fuck* sounds so ugly,
Like playing some dirty trick. That bat's skin
Of mamma's is inside people's heads.
But it's what we grew up with, that red-neck
Way of not respecting a woman or a kid
Or some little animal because life's been too hard

And men have had to work in the sun
Till they hate anything that's weaker than they are.
When my daddy took me hunting, he shot one quail,
Then his dog started flushing them too soon
And he got real mad. When a rabbit jumped up,
He shot it just with his bird load. I found it
Crouched in some briars with its eyes bugged out
Like holding its breath. "You hurt it but it's still
Not dead," I yelled. "It's hurt bad enough,"
He hollered back. I killed it with a rock.
Then I couldn't make myself quit crying.
I had that problem when I was little.
I'd almost quit, then my breath would hiccup,
And the sobbing in my lungs would start all over.

What brought this up was the luck we had
With our lot. The power company did it.
We went out there on a Sunday and they'd cut
This right-of-way from next to the creek.
Their tarry light poles lying there looked rusty
As cannons. The trees were chain-sawed down
So it reminded me of pictures of Civil War battlefields,
People piled up with their shirts blown open.
The stream looked naked in the sun, like some
Young girl that they'd pulled the clothes off of.
Roger really loves me, I guess. His face got
Tender when I kept on sitting on that pine log
With catches hiccuping in my breathing, as if the girl
Who had heard daddy's linemen "mother fucking,"

And their joking sneakiness like greasy mechanics,
Like their hands would smear you in patches like bruises
Wherever they touched, was still inside.
I was talking crazy, said the men fought battles
And tore up cornfields and trees and came home
One-armed and heroic from whichever war
And let women do the worrying. But Roger said
We had us a peace from the old destruction.
He got that real estate man on the phone,
And told him how the woods had been ruined.
That salesman got his mess to the development
And saved the last lot from being cut.
We're going to break ground tomorrow. I dread
That red clay mud and the two-by-fours
With nails sticking through and the leftover
Bags of cement that get hard in the rain,
But some things you have to have to live.
We got tangled up in the briars and the steel-wool
Cedars when we were stringing it off;
Roger got tired of the changes, with me trying
To get it perfect, front toward the street,
And back toward the south for our windows.
But when that little wind cooled our sweat,
And leaves rattled loose here and there,
It was all worthwhile. Those chain-gang masons
In the basement hole can cuss like pure hell,
When sun makes the clay red as fire.
The clotted-up lumps can remind me of folks
Down home, their faces as purple as brickbats,

When they died with a stroke. Me and Roger
Will remember, we've lived through too much now,
With Jimmy and the other little house and the plans
For this one, with changing the front door,
And what kind of linoleum to buy,
And agreeing on fixtures for the front-porch light.
We've kept love together this far.
I know I see windows where the trees are now.
There must be a light in the evenings out there
To brighten up our lives in one place,
With a smooth surface on it like a mirror.
Sometime it has to come as clear as I've dreamed.

ABOUT THE SERIES

Since 1975, Virginia Commonwealth University has sponsored publication, under the general editorship of Walton Beacham, of the winning manuscript in the annual AWP Award Series in Poetry, an open competition for book-length manuscripts. Established in 1974 in a cooperative arrangement between VCU and the University Press of Virginia, the award carries a $500 honorarium and an invitation for the winning author to read at the AWP Annual Meeting.

Manuscripts are received by the series director, who divides them among readers, who are published poets. Finalists are selected and the manuscripts are submitted to a final judge who chooses the winning book. Final judges for the series have included Richard Eberhart, Elizabeth Bishop, Robert Penn Warren, and Maxine Kumin. Donald Justice chose *Following Gravity* as the first place selection in the 1979 AWP Award Series.

For further information and guidelines for submission write: The Associated Writing Programs, Old Dominion University, Norfolk, Virginia 23508.

ABOUT THE AUTHOR

Born (1935) and reared in Stantonsburg, a town of a thousand people in tobacco-growing eastern North Carolina, James Applewhite was educated at Duke University (B.A., M.A., Ph.D.). His first teaching experience was at the University of North Carolina at Greensboro (then Woman's College), where he became acquainted with the poets Randall Jarrell and Robert Watson and, later, Allen Tate. He received a Danforth Fellowship for excellence in teaching, was awarded one of the *Virginia Quarterly Review*'s Emily Clark Balch prizes, and had a poem selected by Denise Levertov for inclusion in the *American Literary Anthology*, vol. 3.

He returned to Duke University in 1972 to teach the writing of poetry and the English Romantic poets and is now an associate professor. In 1975 he received a Creative Writing Fellowship from the National Endowment for the Arts, had his first book of poems published by the University of Georgia Press (*Statues of the Grass*), and traveled with his family in England and France. He was awarded a Guggenheim Fellowship in 1976 and again traveled abroad.

Mr. Applewhite's poems have appeared in the *Borestone Mountain Poetry Awards*, and the selection of his work made by Paul Carroll for his anthology *The Young American Poets* attracted wide attention. He has continued to publish poems in such magazines and quarterlies as *Harper's*, *Esquire*, the *Atlantic*, *Poetry*, *Sewanee Review*, *Virginia Quarterly Review*, *Shenandoah*, *North American Review*, *Michigan Quarterly Review*, and *Poetry Now*. A group of his poems was chosen by Stanley Kunitz for inclusion in a special supplement to the *American Poetry Review*. In 1978 he attended the Bread Loaf Writer's Conference as a John Atherton Fellow. He is a member of the Poetry Society of America.

Mr. Applewhite lives with his wife, Janis, and their three children in northern Durham County, on a small stream that runs into the Eno River. He is an avid jogger and hiker of the neighboring woodland and riverbank, and an amateur carpenter.